Anne S. Bittker
Children's Collection

NEW HAVEN
FREE PUBLIC
LIBRARY

A CIRCLE IS NOT A VALENTINE

H. Werner Zimmermann

TORONTO OXFORD NEW YORK
OXFORD UNIVERSITY PRESS
1990

Ahhh!
It's almost Valentine's Day.
I'm going to paint a Valentine for you.

A Valentine is special.
It says, "I Love You."

I'll show you how it's done.
It's as easy as one, two, three.

Ta-daaa!
There it is. Just for you.

What do you mean, it's not a Valentine?
What is it?

Of course!
It's a square.
How silly of me.

Well then, I'll just have to change it.
For you, it must be perfect.

A circle.
Certainly it's a circle.
I just wanted to see
if you were paying attention.

And now for the Valentine.

It is not a circle.
It is not a square.
And. . . it is not a Valentine.

What is it?

A triangle is not a square.
A square is not a circle.
And a circle is not a Valentine!

How can I show you how I feel?

How will you know what's in my heart?

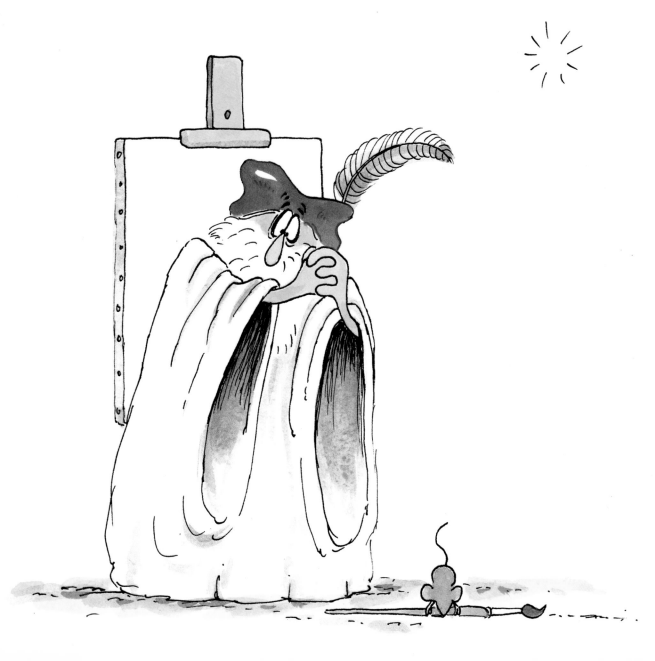

Ahhh!

Oxford University Press, 70 Wynford Drive, Don Mills, Ontario, M3C 1J9

Toronto Oxford New York Delhi Bombay Calcutta Madras Karachi
Petaling Jaya Singapore Hong Kong Tokyo Nairobi Dar es Salaam
Cape Town Melbourne Auckland

and associated companies in
Berlin Ibadan

Canadian Cataloguing in Publication Data
Zimmermann, H. Werner (Heinz Werner), 1951-
Alphonse knows— a circle is not a valentine

ISBN 0-19-540744-X

1. Form perception - Juvenile literature.
I. Title.

BF293.Z55 1990   j152.1'4   C89-094745-7

Oxford is a trademark of Oxford University Press
1 2 3 4 – 3 2 1 0
Printed in Hong Kong